Complete Your Pre-Summer Activity

Visit: lumoslearning.com/book

Enter Access Code: CRX6

OR

Scan QR code

Lumos Summer HeadStart Workbook, 1st to 2nd Grade: Math, Reading, Writing, Science, Fun Activities, Online Resources & AI Tutor.

Contributing Author - Bonnie McRae
Contributing Author - Amber Williams
Contributing Author - Mary Waters Cox
Executive Producer - Mukunda Krishnaswamy
Program Director - Anirudh Agarwal
Designer and Illustrator - Sowmya R

Updated in June 2026

ISBN 13: 979-8897130993

Printed in the United States of America

For permissions and additional information, contact us

Lumos Information Services, LLC
PO Box 1575, Piscataway, NJ 08855-1575
http://www.LumosLearning.com

Email: support@lumoslearning.com
Tel: (732) 384-0146
Fax: (866) 283-6471

Lumos Learning
Step Up Your Skills

Table of Contents

Introduction

Summer is here, which brings a time of excitement, relaxation, and fun. School is the last thing on your mind, but that doesn't mean learning has to be on vacation too. In fact, learning is just as important, and can be just as fun (if not more), during the summer months as during the school year.

Why Is Summer Learning So Important?

Studies show that if students take a standardized test at the end of the school year, and then repeat that test when they return in the fall, they will lose approximately four to six weeks of learning. This loss is commonly referred to as the summer slide. Because of this, it is imperative your children work to keep their minds sharp over the summer.

Taking some time during the summer months to revisit the concepts learned during the school year will help prevent learning loss and help children have a successful back-to-school experience.

What Is Included In This Book?

This book offers various standards-aligned resources, in both printed and online format, to help students learn during the summer months.

The activities in the book are organized by week and aligned with the 1st-grade learning standards. We encourage you to start at the beginning of the Summer holidays. During each week, students can complete daily Math and English practice. There are five daily practice worksheets for each week.

Students can log in to the online program once a week to do math practice, reading, vocabulary and writing activities, and take the science quiz.

Happy Summer Learning!

Week 1 Count and Add

Day 1 - Math

Color the correct total.

1.  +  = 7 / 9 / 11

2.  +  = 10 / 12 / 15

3.  +  = 13 / 11 / 14

Find Your Name

Week 1

Day 1 - English

Find the air balloons with the letters of your name. Color them with your favorite color.

My name is ________________________________

Week 1 Number Train

Day 2 - Math

Count the animals and fill the boxes with the correct numbers

1.

2 + 2 + 1 + 1 = 6

2.

☐ + ☐ + ☐ + ☐ = ☐

3.

☐ + ☐ + ☐ + ☐ = ☐

Count the Nouns

Week 1

Day 2 - English

Read the sentences and count the nouns in each sentence. Color the correct answer

1. Humpty Dumpty sat on a wall.

1 | 2 | 3

2. Jack and Jill went up the hill.

1 | 2 | 3

3. The wheels of the bus go round and round.

1 | 2 | 3

Week 1 Count back to Subtract

Day 3 - Math

Circle the correct number.

15 - 7 = ?

1. 1 2 3 4 5 6 7 8 9 10 11 12 13 14 15 16 17 18 19 20

7   8   9   10

19 - 8 = ?

2. 1 2 3 4 5 6 7 8 9 10 11 12 13 14 15 16 17 18 19 20

11   12   13   10

13 - 3 = ?

3. 1 2 3 4 5 6 7 8 9 10 11 12 13 14 15 16 17 18 19 20

7   8   9   10

Capitalization

Week 1

Day 3 - English

Circle the words that should be capitalized.

kite

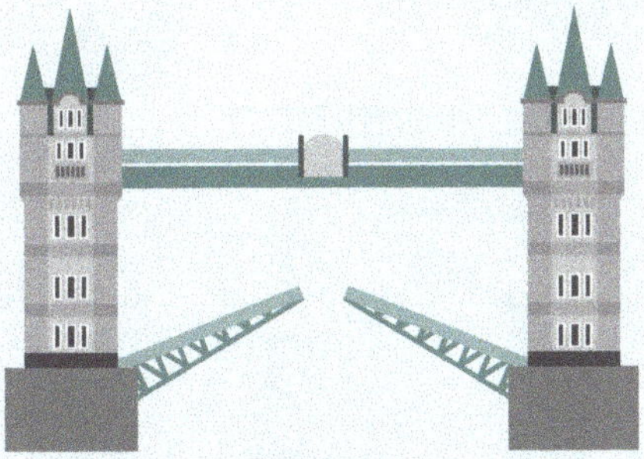

london

bus

rabbit

lisa

miami

coffee

monday

plant

butterfly

Week 1

Adding 3 Digits

Day 4 - Math

Look at the picture and check the correct number sentence.

1. The number of pandas, elephants and lions in the zoo are

3+1+1=5	1+1+2=4	2+2+2=6
☑	☐	☐

2. The number of birds, monkeys and alligators are

3+3+2=8	4+2+1=7	2+1+2=5
☐	☐	☐

3. The number of hippos, pandas, and monkeys are

1+1+3=5	1+2+1=4	1+3+2=6
☐	☐	☐

Day 4 - English

Read the poem

Snowball

by Shel Silverstein

I made myself a snowball
As perfect as could be.
I thought I'd keep it as a pet
And let it sleep with me.
I made it some pajamas
And a pillow for its head.
Then last night it ran away,
But first it wet the bed.

Check the words you see in the poem.

sleep	☐	clap	☐	night	☐	red	☐
head	☐	pet	☐	pillow	☐	bat	☐

Join the dots and color the picture.

Week 1 Candy House

Day 5 - Math

Look at the picture and solve the problems. Color the box having the correct answer.

1. There were 6 creamy cupcakes in the candy house. Angelica ate 2 creamy cupcakes. How many are left?

4	2	3

2. There were 7 chocolate cupcakes in the candy house. John ate 4 chocolate cupcakes. How many are left?

3	1	2

3. There were 5 candies in the candy house. John ate 3 candies. How many candies are left?

1	2	3

Week 1 Word Search

Day 5 - English

Search and circle the words given below.

FRESH	OPEN	SLIP	LET	MORE	DESK

F	R	E	S	H	F	H	S	A
H	U	D	G	N	V	O	L	A
E	F	G	E	S	L	I	P	D
L	I	N	N	D	E	S	K	G
P	D	D	I	C	Q	E	G	H
F	O	P	E	N	W	E	M	M
Z	R	P	G	G	L	E	T	P
X	O	G	I	V	C	J	A	M
M	O	R	E	N	M	Z	R	L

Week 1 Online Activity

- Add & Subtract
- Surprise Guest

lumoslearning.com/a/crx6

Week 2 Smallest Number

Day 1 - Math

Color the smallest number in each row.

1. 

2. 

3. 

Correct the Spellings

Week 2

Day 1 - English

Circle the word that is spelled incorrectly and fill in the blank with the correct spelling of the word.

1. We lik to go fishing. ___________

2. He is not at shool today. ___________

3. When do we go hom after swimming? ___________

4. We hav fun riding our bikes. ___________

5. I love plying basketball. ___________

Week 2 Greatest Number

Day 2 - Math

Color the greatest number in each row.

1.	3	2	5	8
2.	2	9	1	6
3.	7	4	5	3

Punctuation

Week 2

Day 2 - English

Put the correct punctuation at the end of each of the following sentences.

1. Do you like your vacation time
2. That was an amazing goal
3. Where is my new coat
4. He has brown eyes
5. What a great victory

Week 2 Count and Mark

Day 3 - Math

Compare the pictures and mark as <, > or =

1.  

2.  

3. 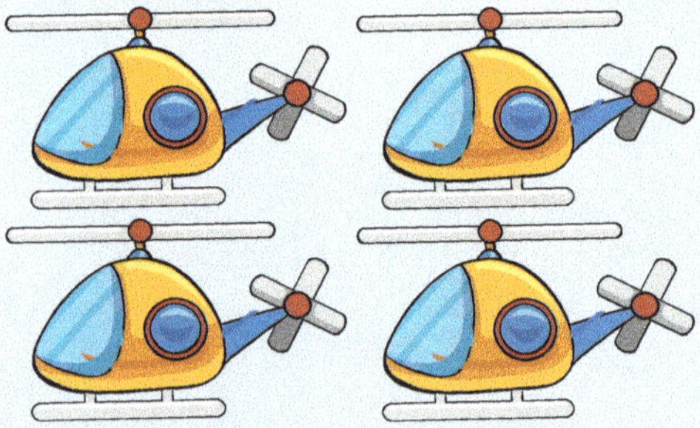 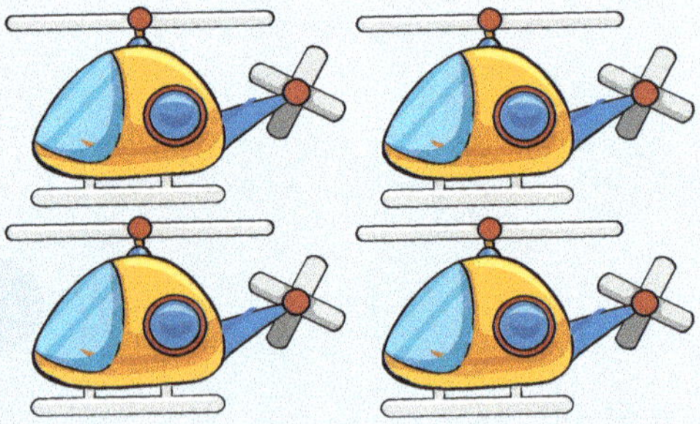

Story Time

Week 2

Day 3 - English

Read the story.

Billy and the Fat Cat

Billy has a fat cat. The cat likes to sleep on Grandma's mat. Grandma does not like that.

Grandma says the fat cat gets hair on her mat.

One day, Billy couldn't find his fat cat, not even on the mat!

"Grandma," he called out. "Have you seen my cat? I cannot find it!"

Grandma was getting ready to go shopping. "No, Billy, I haven't seen it today."

Grandma put on her favorite hat, shoes, and coat to go shopping.

Just then, Billy started to laugh. Grandma asked him, "Why are you laughing and pointing your finger at my hat, Billy? Do I look funny?"

Can you guess where the fat cat was?

Answer the questions based on the story.

1. Where does the fat cat like to sleep?

....

2. Where do you think the cat was?

....

Just for fun!

Draw a picture of a fat CAT wearing a hat and color it.

Week 2 What's Next

Day 4 - Math

What number comes next?

1. 

2	4	6	8	

2. 

3	6	9	12	

3. 

5	10	15	20	

Action Words (Verbs)

Week 2

Day 4 - English

Choose the right action word from the box and write it in the sentence.

helped	slept	swam	fish	hiked

1. My dog, Hattie, __________ up the tall hill and then rested.
2. The dolphin ___________ by our boat.
3. Daniel takes his boat out to ______________ for trout on Saturday.
4. The girl ____________ in the hammock.
5. Howard ______________ in the US Army.

Week 2

Land the Rocket

Day 5 - Math

Skip count by 2 and fill in the missing numbers.

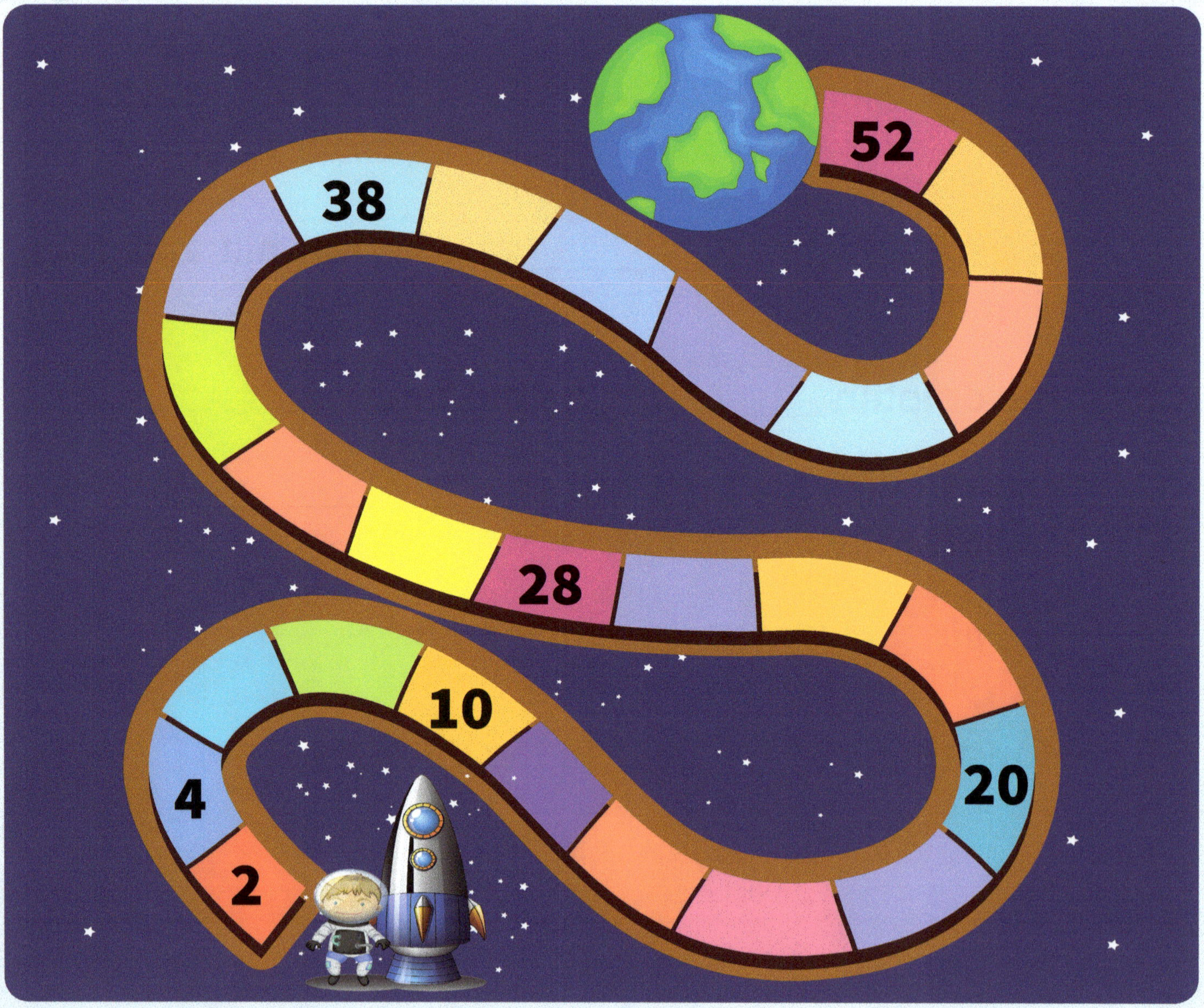

One or Many

Week 2

Day 5 - English

Read the word in each row. Put 'X' in the correct box to show if the word is singular or plural.

Word	Singular	Plural
children		X
cakes		
baby		
fox		
dragons		
lamp		
goose		
horses		
blanket		
cookies		

Week 2 Online Activity

- Number Order
- The Game

lumoslearning.com/a/crx6

Tens and Ones

Week 3

Day 1 - Math

Write the correct tens, ones, and total in the blank places.

Ex.

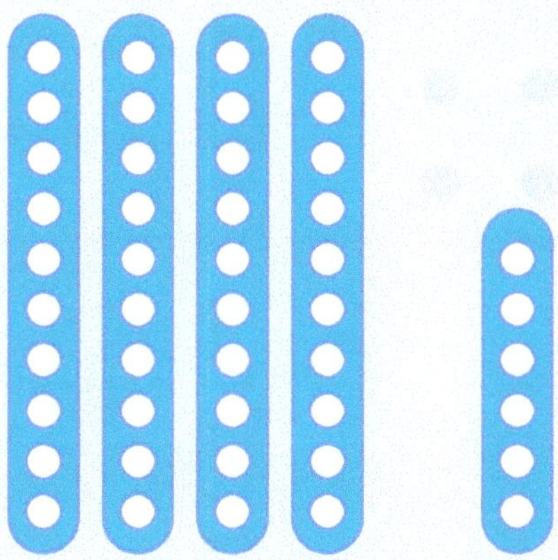

Tens 4 Ones 6

Total 46

1.

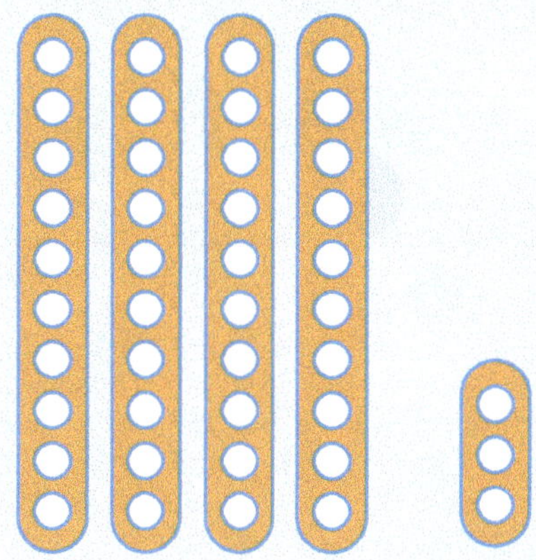

Tens ___ Ones ___

Total ___

2.

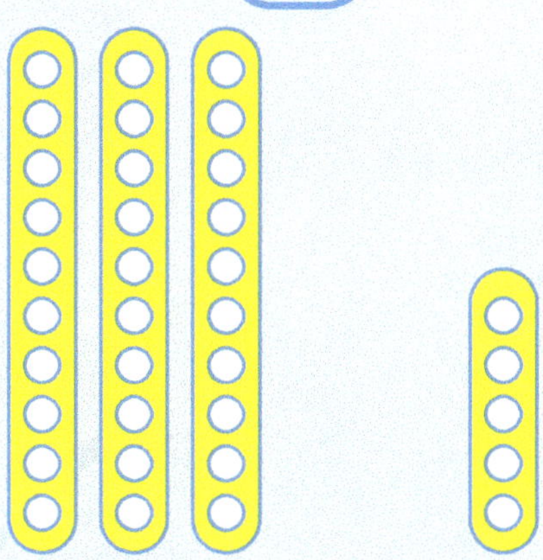

Tens ___ Ones ___

Total ___

3.

Tens ___ Ones ___

Total ___

Week 3 Unscramble

Day 1 - English

Unscramble the letters using the picture clues.

Scrambled	Picture	Unscrambled
kobo		
naaban		
fdoo		
dske		
trckoe		

Even or Odd

Week 3

Day 2 - Math

Count and write the total and Even or Odd in the box below.

		Total	Even/Odd
Ex.		3	Odd
1.			
2.			
3.			

Week 3 Long Vowel Sound

Day 2 - English

Circle the picture that contains the long vowel sound.

ā

ē

ī

ō

ū

Before and After Numbers

Week 3

Day 3 - Math

Write the correct before and after numbers in the blanks.

Ex.

1.

2.

3.

Week 3 Short Vowel Sound

Day 3 - English

Circle the picture that contains the short vowel sound.

ă

ĕ

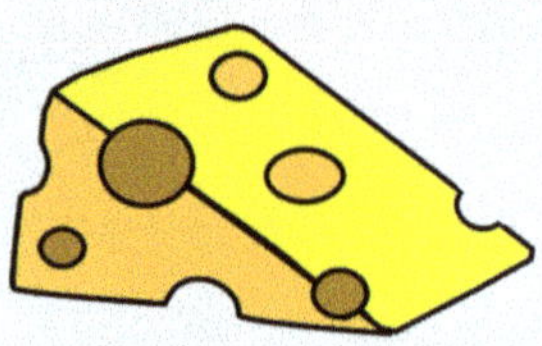

ĭ

ŏ

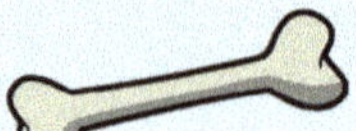

ŭ

Number Order

Week 3

Day 4 - Math

Arrange the numbers in order from lowest to highest.

	Numbers				Order			
Ex.	65	98	24	73	24	65	73	98
1.	63	27	95	32				
2.	46	93	74	59				
3.	82	23	91	47				

Week 3 Poem Time

Day 4 - English

Read the poem below and listen for the rhyming words.

Who Am I?

I live on a farm,
but cause no harm.
I like to make a sound,
Heard from far around.
I can cluck
I am not a duck!
I have two legs,
Yes, I lay eggs!
I am a hen!
A girl chicken!

Find and write the words from the poem that rhyme with each word below.

1. farm ....

2. sound ....

3. eggs ....

4. cluck ....

Just for fun! Draw a picture of a CHICKEN and color it.

Week 3 Hundreds Place

Day 5 - Math

What digit is in the hundreds place in each number?

Ex.	327	3
1.	721	
2.	123	
3.	834	
4.	468	
5.	914	
6.	591	

Sight Words

Week 3

Day 5 - English

Say the words in the box and use each word in the correct sentence. Use each word only once.

your	know	did	help	find

1. He _______ not read the question.

2. Do you ___________ how to multiply?

3. The missing book was hard to ____________.

4. You may ____________ clean the room.

5. It was not her book, it was _____________ book.

Week 3 Online Activity

- Word Problems
- Hare and the Tortoise

lumoslearning.com/a/crx6

Number Words

Week 4

Day 1 - Math

Count the dots and write the numbers in words.

Ex.  eight

1. 

2. 

3. 

4. 

Week 4 Story Time!

Day 1 - English

Read the story

Danny's Pets

Danny has many pets at his house. He has a parrot named Lucky, two cats, Cricket and Socks, and four dogs, Hattie, Scooter, Jax and Lulu.

Danny likes to play with his pets, too. He lets Lucky sit on his arm. He plays with a ball of string with the cats. He takes the dogs for walks around his yard. They like to play chase, too.

He has to feed and water his pets. He feeds his parrot carrots as a treat. His cat likes warm milk. The dogs love chewy bones.

Sometimes he forgets to feed Lucky. Lucky starts to squawk at him. Danny laughs at Lucky. He says, "Sorry, Lucky. I forgot!" Then he feeds him.

Danny thinks taking care of his pets is fun, not work!

Answer the following questions.

1. **What is the setting of the story?**

   (A) The zoo
   (B) Danny's house
   (C) The park
   (D) The yard

2. **What do you think the underlined word "squawks" means in this story?**

   (A) Whistles
   (B) Flaps his wings
   (C) Screeches, or makes loud noises
   (D) Pecks at the cage

Fill in the blanks.

1. **Danny had a ____________________ named Lucky.**
2. **His cat likes ____________________ milk.**
3. **He takes his ____________________ for a walk.**
4. **He feeds ______________________ to his parrot as a treat.**

Just for fun! Draw and color a picture of your favorite pet and give it a name!

Join and Color

Week 4

Day 2 - Math

Connect the dots and color the picture.

Week 4 Complete the Word

Day 2 - English

Look at the pictures and write the first two letters that the word begins with.

_ _uits

_ _uck

_ _hool

_ _ock

Part + Part = Whole

Week 4

Day 3 - Math

Fill the missing part that completes the number 10.

whole 10 → part 8, part 2

A. whole 10 → part 4, part

B. whole 10 → part 9, part

C. whole 10 → part 5, part

D. whole 10 → part 6, part

E. whole 10 → part 3, part

Week 4 Can You Count It?

Day 3 - English

Circle 'C' for countable nouns and 'U' for uncountable nouns.

	Words	Countable	Uncountable
Ex.	Honey	C	U
1.	Book	C	U
2.	Apple	C	U
3.	Sand	C	U
4.	Chicken	C	U
5.	Soup	C	U

Number Balance

Week 4

Day 4 - Math

Fill in the boxes to balance the totals.

Ex. 4 + 4 = 3 + 5

1. 5 + 4 = ☐ + 3

2. 2 + 6 = 4 + ☐

3. 4 + 2 = ☐ + 1

4. 2 + 3 = 5 + ☐

Week 4 I or Me?

Day 4 - English

Complete the sentence by writing I or Me.

1. If you ask __________, this is silly.

2. _________ like to play basketball.

3. The dog followed John and ________ to the door.

4. Will you bake a cake for _________?

5. Jim and _________ went to the carnival.

How Many?

Week 4

Day 5 - Math

Count the similar items and write the number in the box.

Week 4 Compound Words

Day 5 - English

Draw a line from a word on the left to a word on the right to make a compound word.

Ex.	Mail	bye
1.	Sun	mother
2.	Bath	fish
3.	Star	set
4.	Grand	tub
5.	Good	box

Week 4 Online Activity

- Odd and Even
- The Fox and the Goat

lumoslearning.com/a/crx6

Week 5 Find the Odd One

Day 1 - Math

Color the picture that is different.

1. 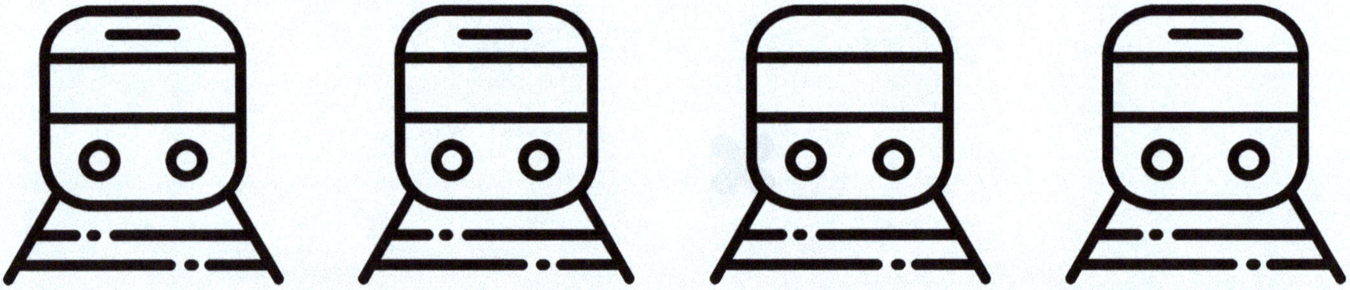

2. 

3. 

4. 

Who is the Narrator?

Week 5

Day 1 - English

Circle the pictures that match the clue.

1. I am a pet. I like milk. I chase mice.

2. I am the king of the jungle. I am very brave.

3. I collect tiny sticks and soft things to make a nest on a tree.

Week 5 Fun with Shapes

Day 2 - Math

Color, trace and draw the shapes.

Color	Trace	Draw

1. 

2. 

3. 

4. 

5. 

Compare the Characters

Week 5

Day 2 - English

Read the Story.

Mary and Susan are sisters.

They are playing on the beach.

Susan is making a sand castle.

Mary is collecting the sea shells on the shore.

Mom has prepared delicious food for them.

Mary wants a sandwich.

Susan wants a pizza.

They are having a fun time at the beach.

Read the sentences. Draw a line to connect each sentence to the character it describes. If the sentence describes both characters, draw lines to both characters.

1. She is on the beach.
2. She is making a sand castle.
3. She is collecting sea shells.
4. She wants a pizza.
5. She wants a sandwich.
6. She is having fun on the beach.

Week 5 Ordinal Numbers

Day 3 - Math

Color the correct picture.

Ex. Color the 4th star.

1. Color the 8th basketball.

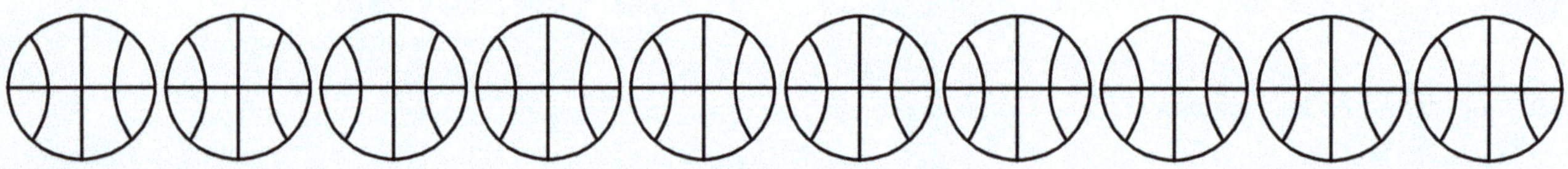

2. Color the 3rd duck.

3. Color the 6th smile.

4. Color the 1st flower.

Common or Proper?

Week 5

Day 3 - English

Check whether the underlined word is a common noun or a proper noun.

	Common	Proper
1. I am going to school.	☐	☐
2. We will visit New York this summer.	☐	☐
3. My dog never stops barking.	☐	☐
4. The bus is stuck in snow.	☐	☐
5. Mark is playing in the field.	☐	☐

Week 5 Addition with 3 Numbers

Day 4 - Math

Add the 3 numbers and write the total in the box.

Ex. 7 + 3 + 4 = 14

1. 6 + 4 + 2 = ☐

2. 2 + 6 + 8 = ☐

3. 5 + 3 + 5 = ☐

4. 1 + 7 + 9 = ☐

Story Time!

Week 5

Day 4 - English

Read the story.

Summer Play

Do you like to play outside in the summer? I do. I am Sam.

My friends, Tony, Mike, Sally, and I play all day if it is sunny. We play baseball, tag, and jump rope. Riding bikes is fun, too.

We always check to make sure we have what we need to play.

We go out after eating breakfast. We come in for lunch or have a picnic in the park.

If it is rainy, we meet at one of our houses. Then we play board games, video games, or watch TV.

All of our parents know where we are and who we are with, so no one gets lost or in trouble.

We listen to our parents' rules and follow them.

It is important to stay together and stay safe. We have great times playing together.

Answer the following questions.

1. What is the story mostly about?

   Ⓐ Rainy days in the summer
   Ⓑ Playing outside and inside with friends in the summer
   Ⓒ Doing chores for our mothers
   Ⓓ My friends' parents

2. What time do Sam and his friends go outside to play?

   Ⓐ After lunch
   Ⓑ Before breakfast
   Ⓒ After breakfast
   Ⓓ When it is dark

3. When do the friends play inside the house?

(A) On a windy day
(B) On a cold day
(C) On a sunny day
(D) On a rainy day

Just for Fun! Draw a picture showing your favorite summer activity and color it.

Week 5 Let's Subtract

Day 5 - Math

Subtract the crossed-out pictures and write the total in the box.

Ex. 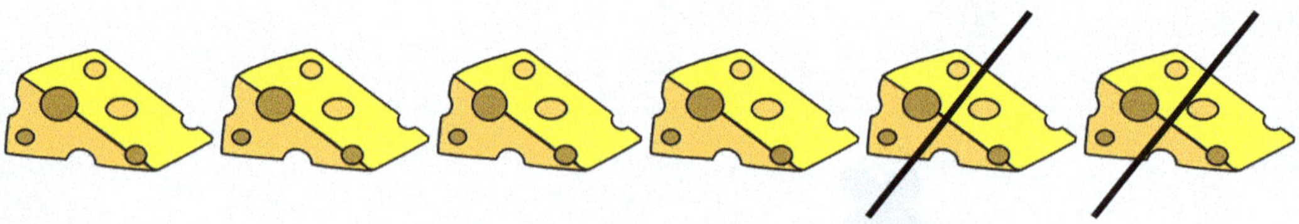 4

1. 

2. 

3. 

4. 

End Blends

Week 5

Day 5 - English

Read and write the words ending in -st, -nt and -lt in the correct boxes.

quilt	twist	hint	tilt
wrist	built	mint	fist
print	spilt	list	tint

-st	-nt	-lt

Week 5 Online Activity

- Fun with Shapes
- Sara Takes a Trip

lumoslearning.com/a/crx6

Count by 10s

Week 6

Day 1 - Math

Count by 10s to fill in the boxes with the missing numbers.

Week 6 Unscramble the Words

Day 1 - English

Read the clues given to Unscramble the Words.

			Clues
Ex.	amp	Map	I need this to find buried treasure.
1.	tca		It loves to catch mice.
2.	dre		An apple is usually this color.
3.	sye		The opposite of no.
4.	heett		We brush them every day.

Add or Subtract

Week 6

Day 2 - Math

Fill in the boxes with + or -

Ex.	5	+	2	=	7
1.	3		1	=	4
2.	6		3	=	3
3.	3		6	=	9
4.	4		2	=	2

Week 6 -ing Words

Day 2 - English

Help Jack to end the words with -ing.

Ex.	sleep	sleeping
1.	eat	
2.	play	
3.	sing	
4.	walk	
5.	think	

Magic Triangles

Week 6

Day 3 - Math

Fill the missing numbers in each triangle so that all three sides add up to the same total.

Ex.
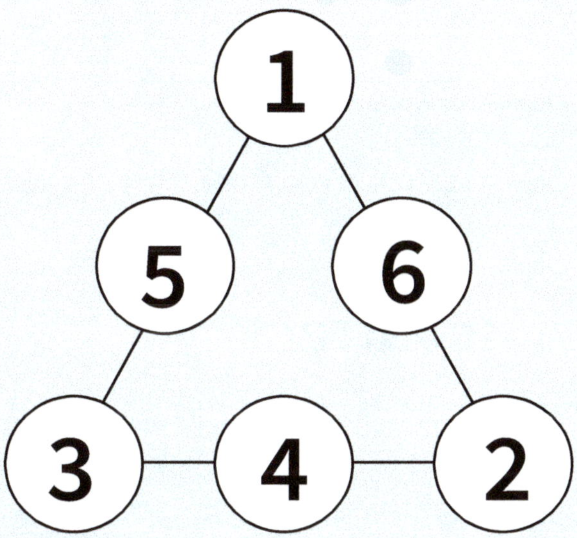

1.
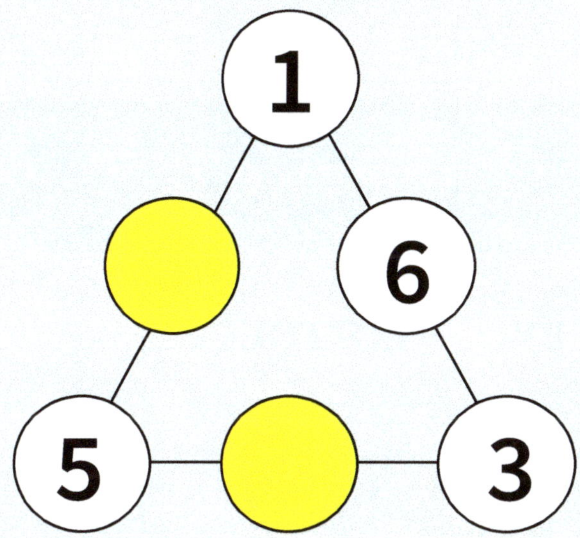

2.
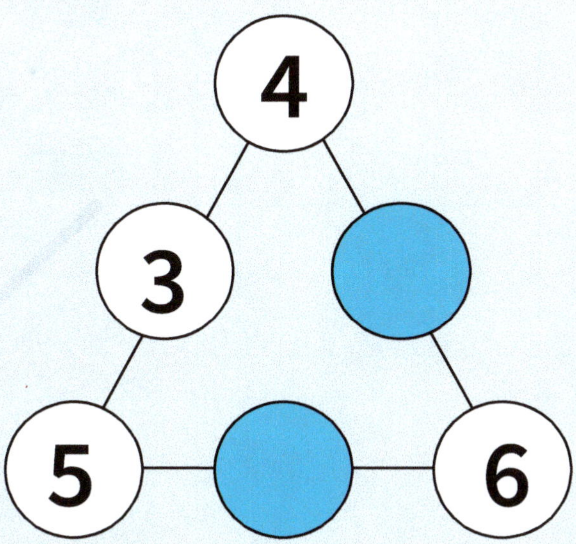

3.
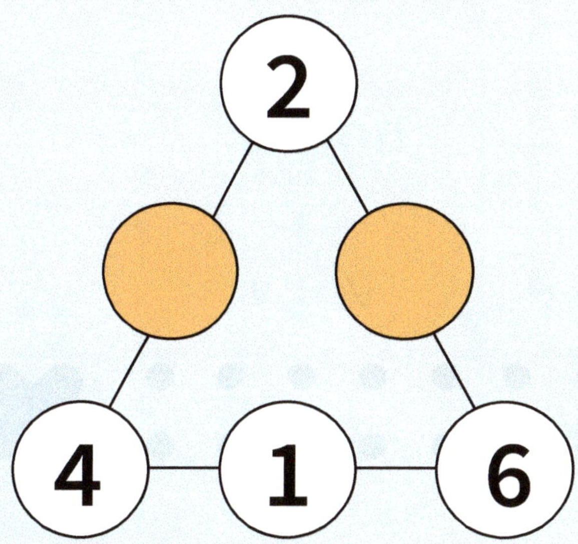

Week 6 Add the Prefix

Day 3 - English

Select the correct prefix from the box and add it to a word below. Use each prefix only once.

A prefix is added to the beginning of a word.

un-	in-	dis-	re-	pre-

	Root word	Prefix	New word
Ex.	correct	in	incorrect
1.	cycle		
2.	lucky		
3.	honest		
4.	view		
5.	side		

Number Maze

Week 6

Day 4 - Math

Follow the correct number sequence to find the treasure.

Week 6 Add the Suffix

Day 4 - English

Select the correct suffix from the box and add it to a word below. Use each suffix only once.

A suffix is added to the end of a word.

-ly	-ful	-ed	-less	-er

	Root word	Suffix	New word
Ex.	Care	ful	Careful
1.	Cheer		
2.	Jump		
3.	Friend		
4.	Strong		
5.	Fear		

Balancing Equation

Week 6

Day 5 - Math

Fill in the number to balance the equation.

Ex. 5 − [3] = 8 − 6

1. 9 − 7 = [ ] − 3

2. 7 − [ ] = 8 − 3

3. [ ] − 5 = 6 − 2

4. 9 − 4 = [ ] − 2

Week 6 Story Time!

Day 5 - English

Read the story.

Timmy and Joey Go Camping

Timmy and Joey are friends. They like to go camping.

The boys take a tent with them. Joey's dad goes, too. He will help set up the tent and start the fire. He takes his flashlight, too.

The boys pack food in an ice chest. Their moms put a first aid kit in the truck.

They each take a sleeping bag, blanket, and pillow.

The boys take clothes that are good for cool and warm weather.

The woods are fun to hike in with Joey's dad. They see squirrels and many plants.

At night, the boys help cook hot dogs and roast marshmallows.

Before they go to bed, they tell silly stories and use a flashlight to shine under their chins.

Find the following words about camping and circle them in the word search given below.

blanket	tent	hike	woods	hot dogs

h	l	m	a	t	u	o	p
i	o	t	e	n	t	x	k
k	i	t	e	m	n	v	o
e	f	l	w	o	o	d	s
a	b	c	g	f	w	o	m
h	o	t	d	o	g	s	i
b	a	n	k	w	q	o	l
b	l	a	n	k	e	t	e

Just for fun! Draw and color 3 items you take with you while camping in the woods.

Week 6 Online Activity

- Tens and Ones
- Mimi and Her Spoons

lumoslearning.com/a/crx6

Week 7 Pattern

Day 1 - Math

Trace and Color the shape that comes next.

Ex.

1.

2.

3.

4.

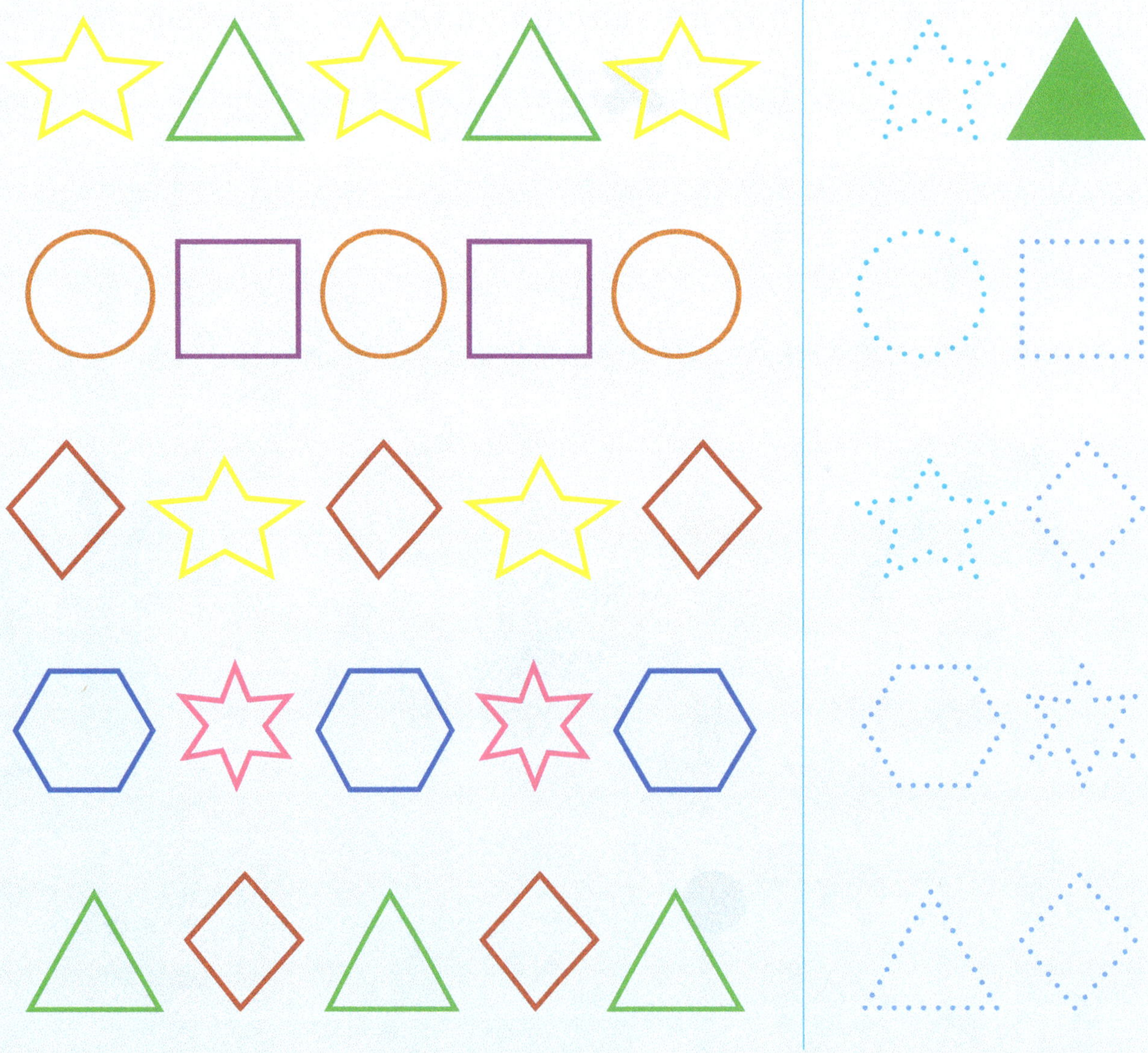

Mixed Up Sentences

Week 7

Day 1 - English

Unscramble the sentences given below.

Ex. has dog a wonderful friend my

My friend has a wonderful dog.

1. years five old is he

2. found on they her the street

3. name is Peter my

4. school i to going love

Week 7 Match the Time

Day 2 - Math

Match the time!!

1. 

2. 

3. 

4. 

3:00
8:00
5:00
12:00

What's the Sound?

Week 7

Day 2 - English

Say the name of the picture. Write the beginning sound of the picture.

C

Week 7 Complete and Color

Day 3 - Math

Draw the other half of the picture.
Color the picture.

Summer Things

Week 7

Day 3 - English

Identify the summer things below.

_a_

f_i_b_e

s_d_

s_r_i_g

_o_l

Week 7

Time Telling

Day 4 - Math

Write the time shown on the clock.

Ex. **9:30**

1.

2.

3.

4.

Story Time!

Week 7

Day 4 - English

Read the Story

Owl on the Tree

There is a tall tree in Mary's garden.
An owl lives in the tree.
Mary can see the owl from her window.
The owl has big, round eyes.
Mary likes to hear the owl hoot.
The owl hunts at night.

Answer the following questions.

1. Where is the tall tree?

....

2. When does the owl hunt?

....

Join the dots and color the picture!

Color by Number

Week 7

Day 5 - Math

Color the picture by number.

1 - Black	2 - Green	3 - Yellow	4 - Orange	5 - Blue
6 - Purple	7 - Brown	8 - Pink	9 - Red	

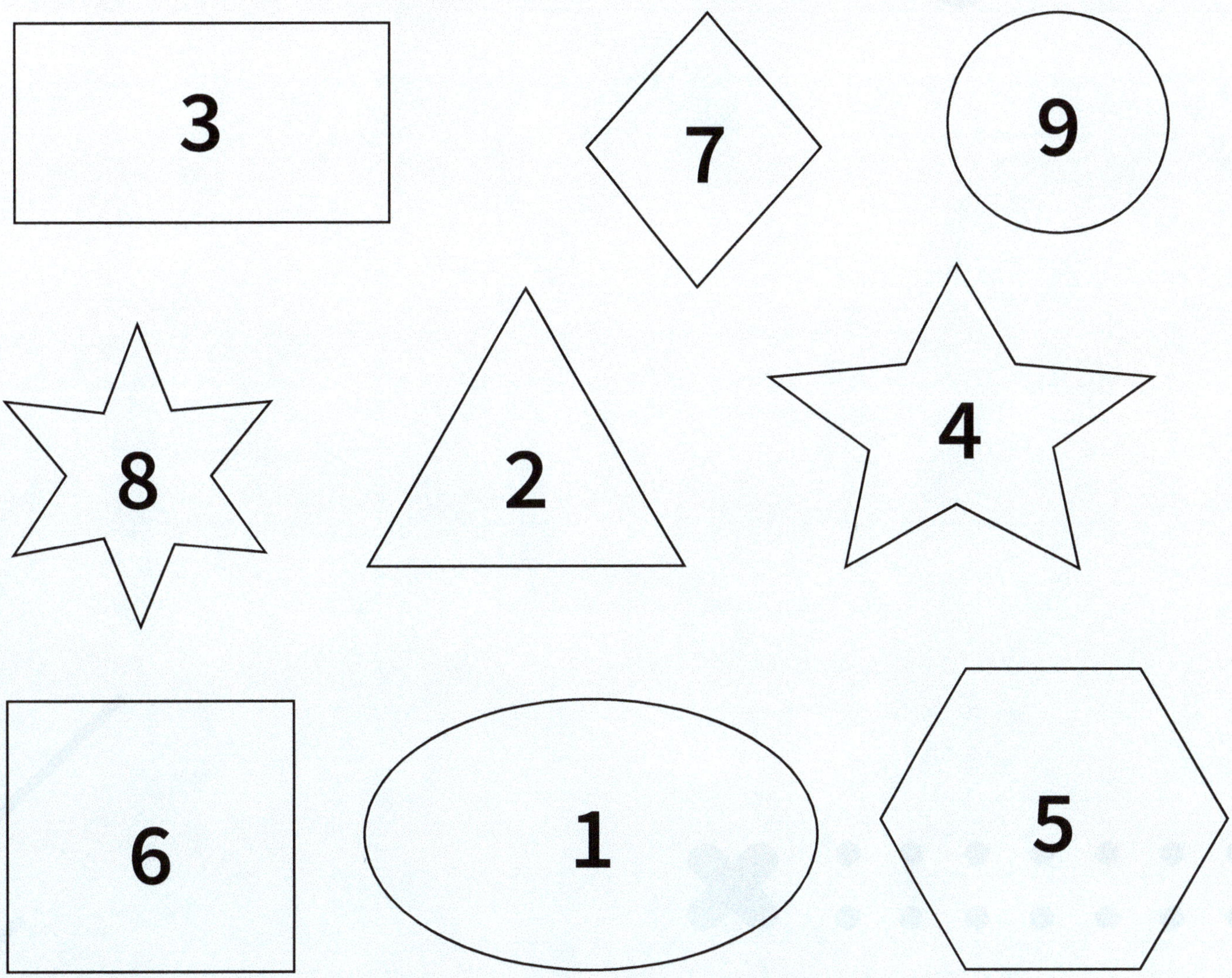

Week 7 Fly, Swim or Walk

Day 5 - English

Sort the words from the box into the correct category.

owl	shark	dog
dolphin	deer	plane
rabbit	bee	whale

FLY	SWIM	WALK

Week 7
Online Activity

- Number Pattern
- Javier and His Friends Fly a Kite

lumoslearning.com/a/crx6

Week 8 Longer or Shorter

Day 1 - Math

Answer the questions by circling the correct picture.

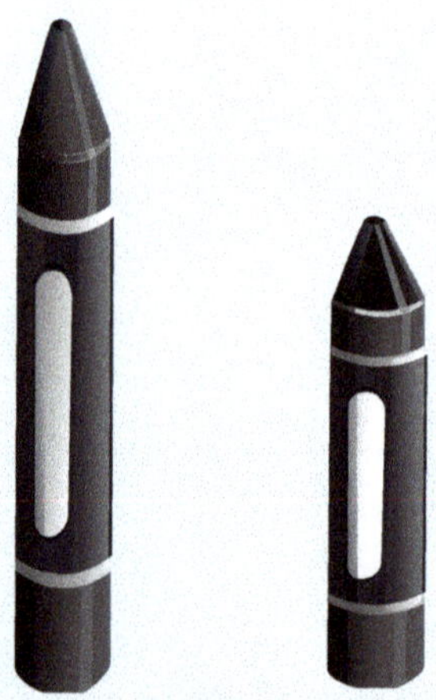

Which is shorter?

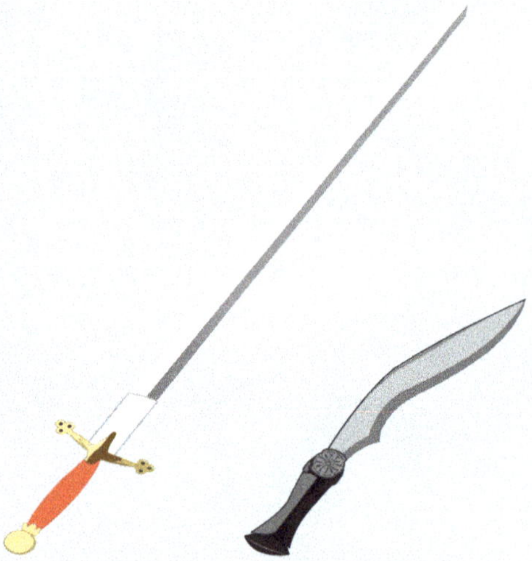

Which is longer?

Which is longer?

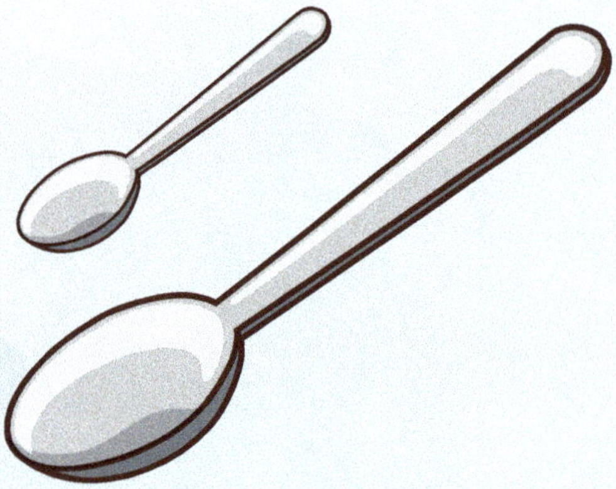

Which is shorter?

About Me

Week 8

Day 1 - English

Tell us about yourself.

1. My name is ______________________________.
2. I am ______________ years old.
3. I like to play ______________________________.
4. I like to eat ______________________________.
5. My favorite color is ______________________.

Draw a picture of yourself and color it!

Week 8 Measuring Length

Day 2 - Math

Using a ruler, measure the length of the following pencils in centimeters.

Ex.

17 cm

1.

______ cm

2.

______ cm

3.

______ cm

A Beautiful Day!

Week 8

Day 2 - English

Write the sentences. Start with a capital letter. End with a period.

1. the sun is shining brightly

2. the grass is green

3. the birds are chirping

4. we are going on a picnic

Week 8 Heavier or Lighter

Day 3 - Math

Answer the questions by circling the correct picture.

Which object weighs less?

Which object weighs more?

Which object weighs more?

Which object weighs less?

Which object weighs less?

Which object weighs more?

Spell - Well

Week 8

Day 3 - English

Color the picture and circle the correctly spelled word.

	pet pat pett
	be bee bea
	rose rows roas
	elephant elepant elaphent

Week 8 Count and Tally

Day 4 - Math

Count the pictures and make tally marks, as shown below.

	Count	Tally	Number
		𝍸 |	6
1.			
2.			
3.			
4.			
5.			

Number the Story

Week 8

Day 4 - English

Read the sentences and write the numbers 1, 2 and 3 in the box to form a short story.

Ex.	Meg tried on three pairs of shoes.	2
	Meg's Mom took her to the shoe store.	1
	Meg liked her new shoes.	3

a.	Joe went fishing with his dad.	
	Joe's family had fish for dinner.	
	Joe caught a fish.	

b.	Jane's teacher said the bird was pretty.	
	Jane gave her picture to her teacher.	
	Jane drew a picture of a bird.	

Week 8

Divide the Shape

Day 5 - Math

Draw a line for each shape to divide it into 2 equal parts.

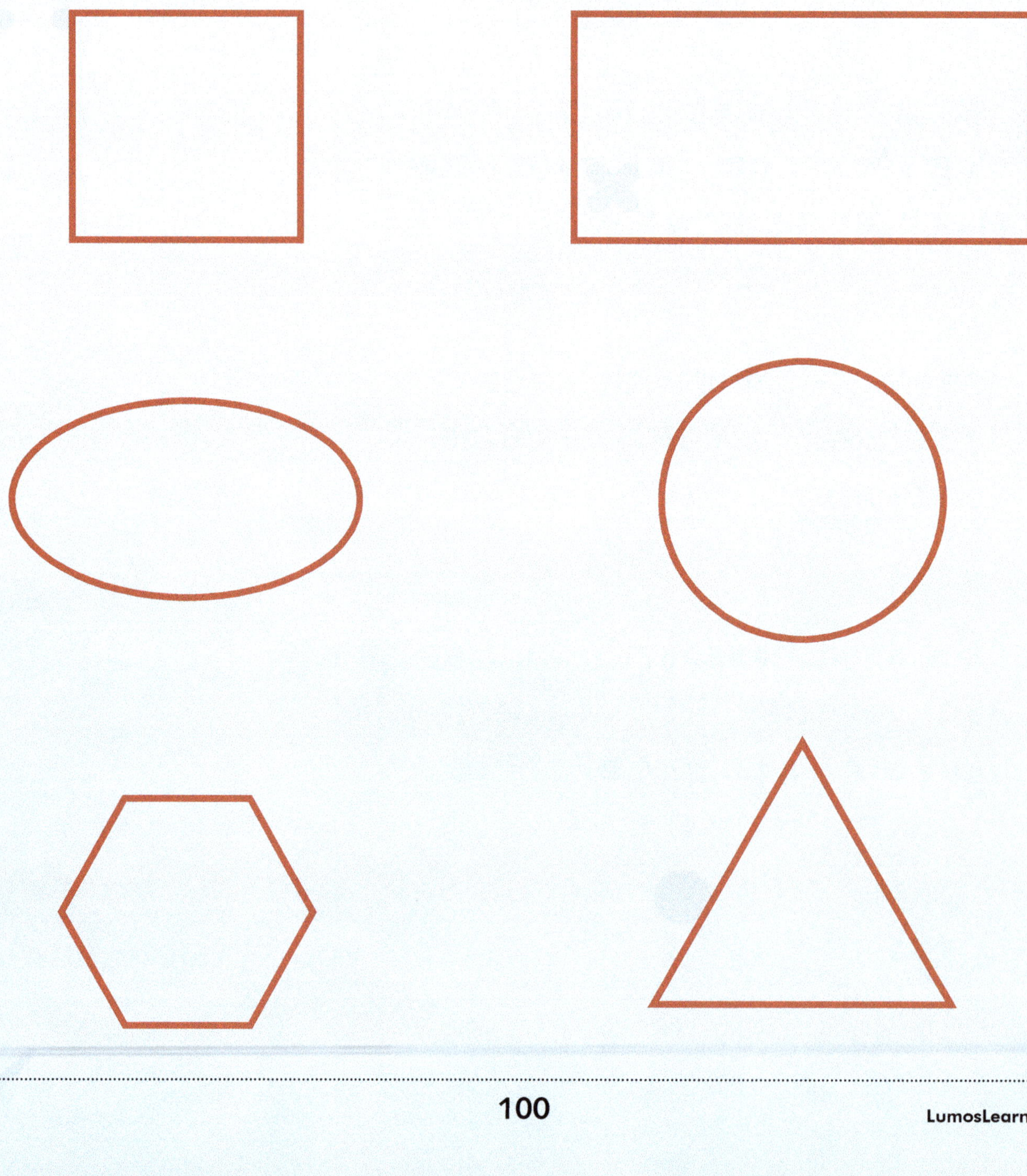

To, Too or Two

Week 8

Day 5 - English

Complete the sentences with to, too, or two.

1. I am going ______ the park at ________ o'clock.

2. I'm ________ young ______ go fishing alone.

3. I have ________ dollars in my piggy bank.

4. My dog is ________ fat.

Week 8 Online Activity

- Measurement
- Chloe Finds a Baby Chick

lumoslearning.com/a/crx6

Color the Shapes

Week 9

Day 1 - Math

Color the fraction shown in each shape.

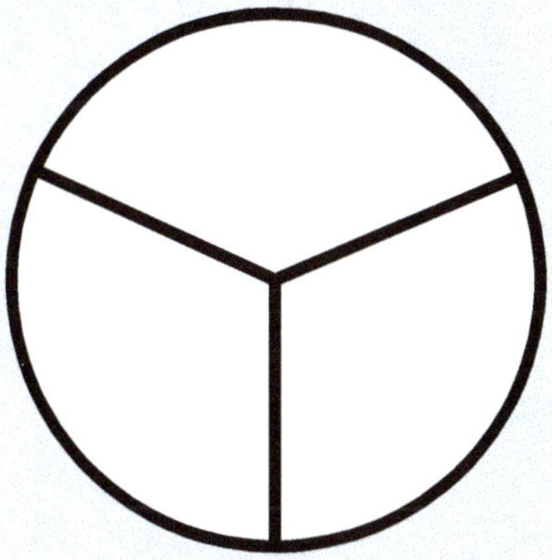

$\frac{1}{3}$

$\frac{6}{8}$

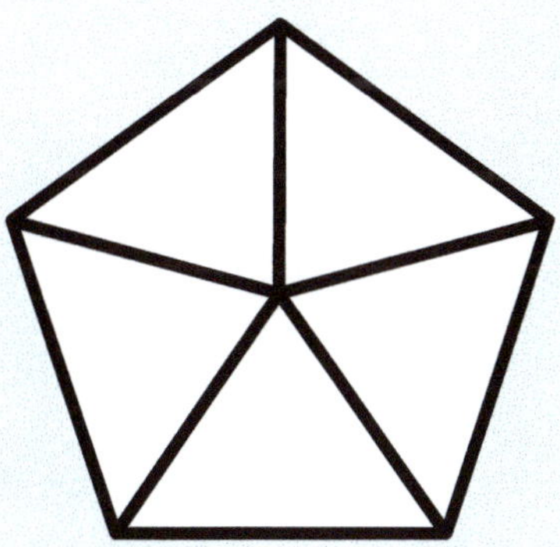

$\frac{3}{5}$

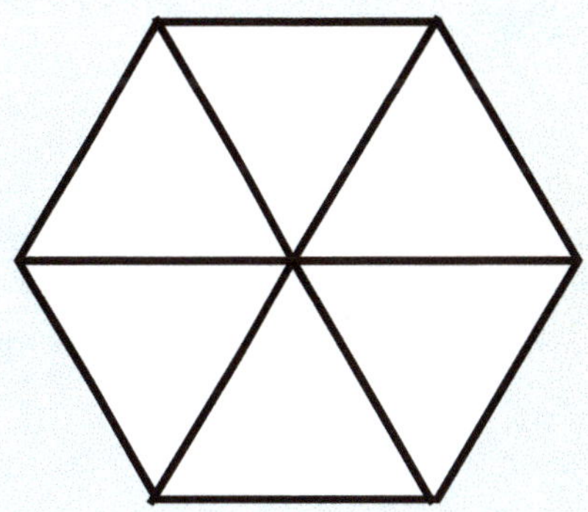

$\frac{2}{6}$

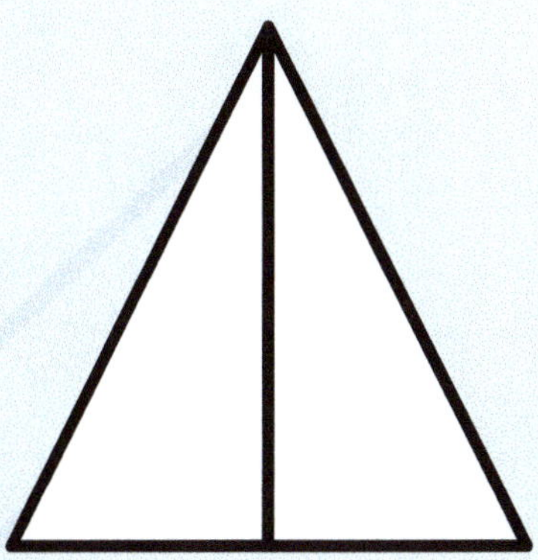

$\frac{1}{2}$

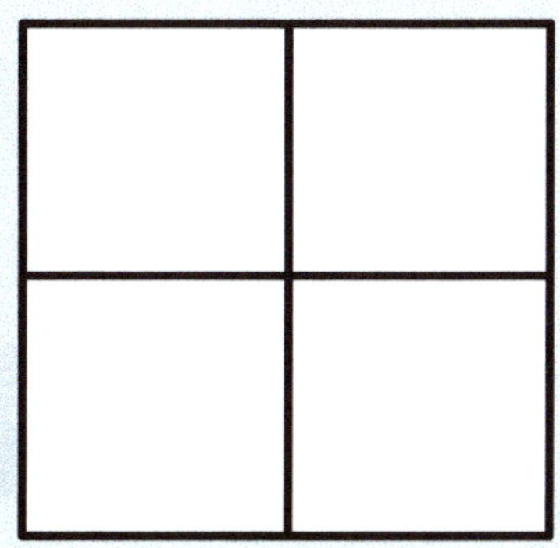

$\frac{2}{4}$

Week 9 Adjectives

Day 1 - English

Choose the best adjective from the box for each picture.

hot	angry	happy	sad	funny

Feeling ______________

Feeling ______________

Feeling ______________

Feeling ______________

Feeling ______________

Animal Bar Graph

Week 9

Day 2 - Math

Create a bar graph.

Pet Animal	Number of Votes
	5
	3
	4
	2

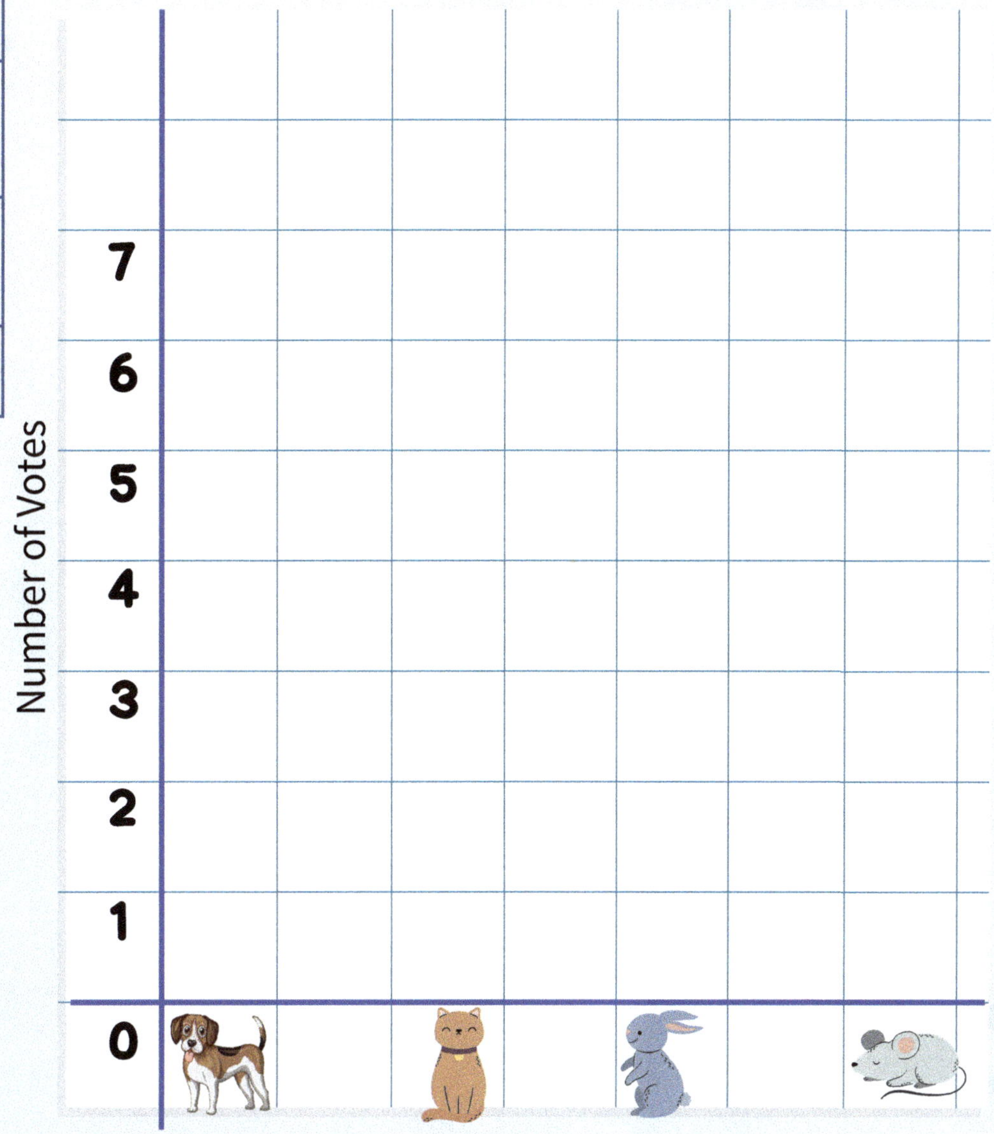

Week 9 Group the Words

Day 2 - English

Cross out the word that does not belong to the group.

1. cap, winter, spring, summer

2. hen, eagle, rabbit, owl

3. pencil, eraser, ruler, basketball

4. ear, nose, eyes, banana

Shape inside a Shape

Week 9

Day 3 - Math

Draw the shapes and color them.

Draw a circle inside a triangle.	Draw a square inside a circle.
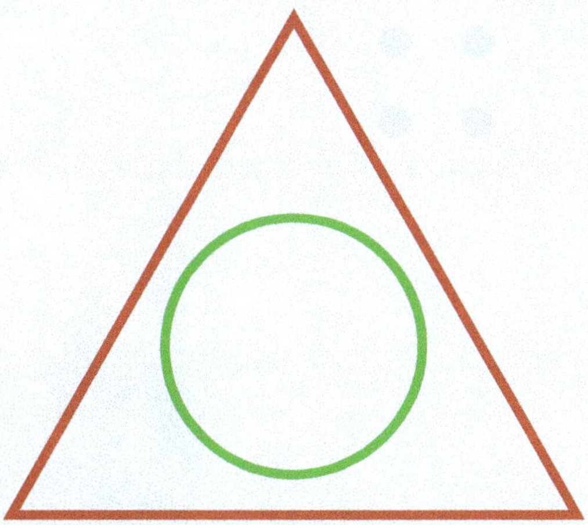	
Draw a circle inside a square.	Draw a square inside a triangle.

Week 9 Picture Match

Day 3 - English

Draw a line from each sentence to its matching picture.

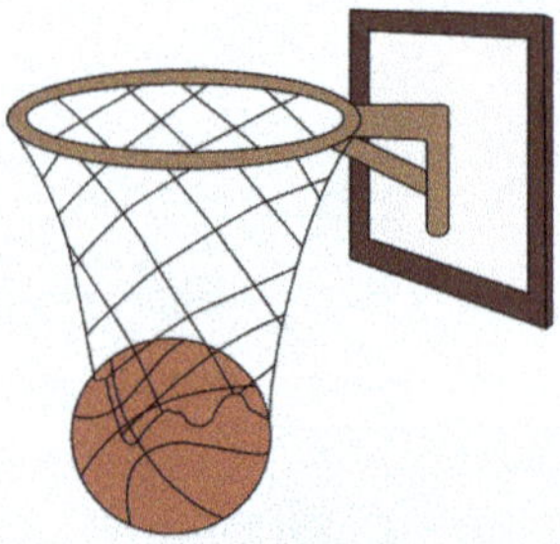

1. This is my dog Pluto.

2. There are four flowers.

3. I love ice-cream.

4. I like playing basketball.

Fast or Slow

Week 9

Day 4 - Math

Circle the activity which takes longer.

Painting a house
or
brushing teeth

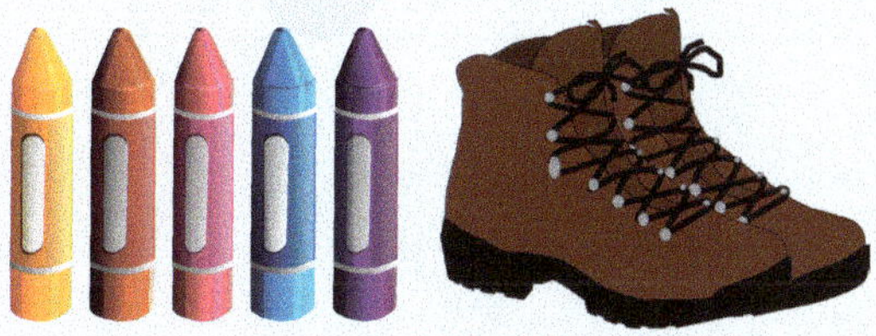

Coloring a picture
or
putting on shoes

Putting on a jacket
or
eating a burger

Playing basketball
or
eating snacks

Week 9

A or An

Day 4 - English

Circle 'a' or 'an' to form the correct sentence.

1. I saw a / an owl on the tree.

2. I took a / an umbrella to the school.

3. We were sitting on a / an bench.

4. Dad read us a / an story at bedtime.

5. A / An apple a day keeps the doctor away.

Draw the Time

Week 9

Day 5 - Math

Draw the hour and minute hand on the clock to show the given time.

6:30

3:30

9:00

4:00

12:30

1:30

8:00

2:30

5:00

Week 9 Story Time!

Day 5 - English

Read the Story.

The Bicycle

Emily has a new bicycle. It is red and shiny. It was a gift from her aunt. She hid it behind a bush to surprise her. When Emily looked behind the bush, she saw the bicycle and jumped with joy. It was just what she wanted. She gave her aunt a big hug. She loves her new bicycle and rides it every day.

Circle the correct answer.

1. What color is the bicycle?

   Pink Red Blue

2. Who gave Emily the bicycle?

   Aunt Uncle Dad

3. Where was the bicycle hidden?

   In her room In the garage Behind the bush

Color the picture with your favorite bicycle color.

Week 9
Online Activity

- Time and Money
- Twins Mac And Madi's Birthday

lumoslearning.com/a/crx6

Answer Key

Week 1

	Lesson Name	Answer Key
Day 1	Count and Add	1) 9 2)10 3)13
	Find Your Name	Answer may vary.
Day 2	Number Train	1) 6 2) 6 3) 10
	Count the Nouns	1) 2 2) 3 3) 2
Day 3	Count back to Subtract	1) 8 2) 11 3) 10
	Capitalization	London, Miami, Lisa, Monday
Day 4	Adding 3 Digits	1) 5 2) 7 3) 6
	Poem Time! - Snowball	sleep, pet, night, pillow, head
Day 5	Candy House	1) 4 creamy cupcakes 2) 3 chocolate cupcakes 3) 2 candies
	Word Search	F R E S H F H S A H U D G N V O L A E F G E S L I P D L I N N D E S K G P D D I C Q E G H F O P E N W E M M Z R P G G L E T P X O G I V C J A M M O R E N M Z R L

Week 2

	Lesson Name	Answer Key
Day 1	Smallest Number	1) 4 2)1 3)2
	Correct the Spellings	1) like 2) school 3) home 4) have 5) playing
Day 2	Greatest Number	1) 8 2) 9 3) 7
	Punctuation	1) Do you like your vacation time? 2) We are going to a Giants's game! 3) Where is my new coat? 4) He has brown eyes. 5) What a great victory!
Day 3	Count and Mark	1) > 2) < 3) =
	Story Time! - Billy and the Fat Cat	1) 2)
Day 4	What's Next	1) 10 2) 15 3) 25
	Action Words (Verbs)	1) hiked 2) swam 3) fish 4) slept 5) helped
Day 5	Land the Rocket	**2**, **4**, 6, 8, **10**, 12, 14, 16, 18, **20**, 22, 24, 26, **28**, 30, 32, 34, 36, **38**, 40, 42, 44, 46, 48, 50, **52**
	One or Many	**Singular**: baby, fox, lamp, goose, blanket **Plural**: cakes, dragons, horses, cookies

Week 3

	Lesson Name	Answer Key
Day 1	Tens and Ones	1)Tens = 4, Ones = 3, Total = 43 2)Tens = 3, Ones = 5, Total = 35 3)Tens = 2, Ones = 4, Total = 24
	Unscramble	kobo = book naaban = banana fdoo = food dske = desk trckoe = rocket
Day 2	Even or Odd	1) 4 - Even 2) 5 - Odd 3) 6 - Even
	Long Vowel Sound	a = train e = tree i = kite o = goat u = flute
Day 3	Before and After Numbers	1)37, 38, 39 2)48, 49, 50 3)82, 83, 84
	Short Vowel Sound	a = cap e = bell i = ship o = sock u = duck
Day 4	Number Order	1) 27, 32, 63, 95 2) 46, 59, 74, 93 3) 23, 47, 82, 91
	Poem Time! - Who am I?	1)farm - harm 2)sound - around 3)eggs - legs 4) cluck - duck
Day 5	Hundreds Place	1)721 - 7 2)123 - 1 3)834 - 8 4)468 - 4 5)914 - 9 6)591 - 5
	Sight Words	1) did 2) know 3) find 4) help 5) your

Week 4

	Lesson Name	Answer Key
Day 1	Number Words	1)five 2)seven 3)six 4)ten
	Story Time! - Danny's Pets	1) Option B 2) Option C Fill in the blanks answers: 1) parrot 2) warm 3) dogs 4) carrots
Day 2	Join and Color	Fun activity
	Complete the Word	**Fr**uits **Tr**uck **Sc**hool **Cl**ock
Day 3	Part+Part=Whole	A)6 B)1 C)5 D)4 E)7
	Can You Count It?	1)Book - Countable 2) Apple - Countable 3) Sand - Uncountable 4) Chicken - Countable 5) Soup - Uncountable
Day 4	Number Balance	1) 5 + 4 = **6** + 3 2) 2 + 6 = 4 + **4** 3) 4 + 2 = **5** + 1 4) 2 + 3 = 5 + **0**
	I or ME	1) me 2) I 3) me 4) me 5) I
Day 5	How Many?	Apples - 4 Orange - 2 Grapes - 3 Cherry - 7 Carrot - 6
	Compound Words	1) Sun + set = Sunset 2) Bath + tub = Bathtub 3) Star + fish = Starfish 4) Grand + mother = Grandmother 5) Good + bye = Goodbye

Week 5

	Lesson Name	Answer Key
Day 1	Find the Odd One	1) 3rd 2)2nd 3)1st 4)4th
	Who is the Narrator?	1) Cat 2) Lion 3) Parrot
Day 2	Fun with Shapes	Math activity
	Compare the Characters	1. Mary and Susan 2. Susan 3. Mary 4. Susan 5. Mary 6. Mary and Susan
Day 3	Ordinal Numbers	Math activity
	Common or Proper?	1) school - Common 2) New York - Proper 3) dog - Common 4) bus - Common 5) Mark - Proper
Day 4	Addition with 3 Numbers	1) 6 + 4 + 2 = 12 2) 2 + 6 + 8 = 16 3) 5 + 3 + 5 = 13 4) 1 + 7 + 9 = 17
	Story Time! - Summer Play	1) Option B 2) Option C 3) Option D
Day 5	Let's Subtract	1) 7 - 2 = 5 2) 8 - 5 = 3 3) 9 - 5 = 4 4) 10 - 4 = 6
	End Blends	-st = twist, wrist, fist, list -nt = hint, mint, print, tint -lt = quilt, tilt, built, spilt

Week 6

	Lesson Name	Answer Key
Day 1	Count by 10s	10, 20, 30, **40**, **50**, **60**, 70, **80**, 90, **100**
	Unscramble the Words	1) Cat 2) Red 3) Yes 4) Teeth
Day 2	Add or Subtract	1) 3 **+** 1 = 4 2) 6 **-** 3 = 3 3) 3 **+** 6 = 9 4) 4 **-** 2 = 2
	-ing Words	1) eating 2) playing 3) singing 4) dancing 5) thinking
Day 3	Magic Triangles	1) 1 + 6 + 3 = 5 + **4** + 1 = 5 + **2** +3 2) 5 + 3 + 4 = 5 + **1** + 6 = 4 + **2** + 6 3) 4 + 1 + 6 = 4 + **5** + 2 = 2 + **3** + 6
	Add the Prefix	1) Recycle 2) Unlucky 3) Dishonest 4) Review 5) Inside
Day 4	Number Maze	Math activity
	Add the Suffix	1) Cheerful 2) Jumped 3) Friendly 4) Stronger 5) Fearless
Day 5	Balancing Equation	1) 9 - 7 = **5** - 3 2) 7 **- 2** = 8 - 3 3) **9** - 5 = 6 - 2 4) 9 - 4 = **7** - 2
	Story Time! - Timmy and Joey Go Camping	h l m a t u o p i o t e n t x k k i t e m n v o e f l w o o d s a b c g f w o m h o t d o g s i b a n k w q o l b l a n k e t e

Week 7

	Lesson Name	Answer Key
Day 1	Pattern	
	Mixed Up Sentences	1) He is five years old. 2) They found her on the street. 3) My name is Peter. 4) I love going to school.
Day 2	Match the Time	1 - 12:00 2 - 5:00 3 - 8:00 4 - 3:00
	What's the Sound?	Tree - t Duck - d Bag - b Grapes - g Hat - h
Day 3	Complete and Color	Fun activity
	Summer Things	Cap, Frisbee, Soda, Surfing, Pool
Day 4	Time Telling	1) 6:30 2) 12:00 3) 3:30 4)12:30
	Story Time! - Owl on the Tree	1) The tall tree is in Mary's garden. 2) The owl hunts at night.
Day 5	Color by Number	Math activity
	Fly, Swim or Walk	Fly = Owl, Plane, Bee Swim = Shark, Dolphin, Whale Walk = Dog, Deer, Rabbit

Week 8

	Lesson Name	Answer Key
Day 1	Longer or Shorter	Math activity
	About Me	Answer may vary
Day 2	Measuring Length	1) 13 cm 2) 11 cm 3) 20 cm
	A Beautiful Day!	1) The sun is shining brightly. 2) The grass is green. 3) The birds are chirping. 4) We are going on a picnic.
Day 3	Heavier or Lighter	Math activity
	Spell - Well	pet bee rose elephant
Day 4	Count and Tally	1) 5 2) 10 3) 7 4) 8 5) 9
	Number the Story	a) 1, 3, 2 b) 3, 2, 1
Day 5	Divide the Shape	Math activity
	To, Too or Two	1. I am going **to** the park at **two** o'clock. 2. I'm **too** young **to** go fishing alone. 3. I have **two** dollars in my piggy bank. 4. My dog is **too** fat.

Week 9

	Lesson Name	Answer Key
Day 1	Color the Shapes	Math activity
	Adjectives	Feeling happy Feeling funny Feeling hot Feeling angry Feeling sad
Day 2	Animal Bar Graph	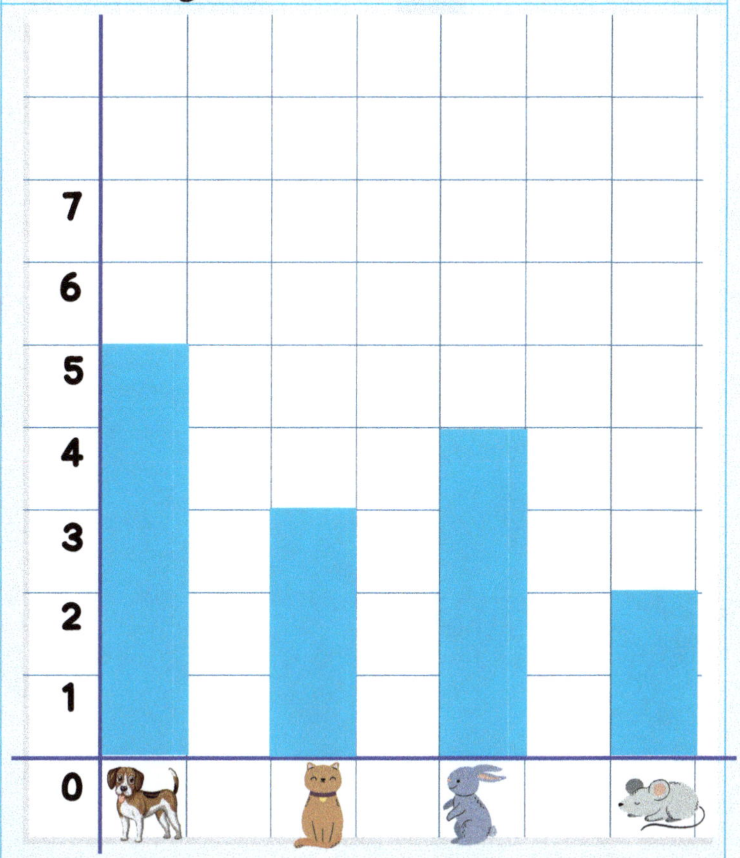
	Group the Words	1) Cap 2) Rabbit 3) Basketball 4) Banana

	Lesson Name	Answer Key
Day 3	Shape Inside a Shape	Draw a square inside a circle:  Draw a circle inside a square:  Draw a square inside a triangle: 
	Picture Match	Fun activity
Day 4	Fast or Slow	Painting a house Coloring a picture Eating Burger Playing basketball
	A or An	1) I saw **an** owl on the tree. 2) I took **an** umbrella to the school. 3) We were sitting on **a** bench. 4) Dad read us **a** story at bedtime. 5) **An** apple a day keeps the doctor away.
Day 5	Draw the Time	3:30  9:00  4:00  12:30  1:30  8:00  2:30  5:00 
	Story Time! - The Bicycle	1) Red 2) Aunt 3) Behind the bush

What if I buy more than one Lumos Study Program?

Step 1 ⟶ **Visit the URL given below and login to your parent account**

www.lumoslearning.com

Step 2 ⟶ Click on the horizontal lines (☰) in the top right-hand corner of the parent account and select **"My tedBooks"**

Place the Book Access Code and submit (See the first page for access code).

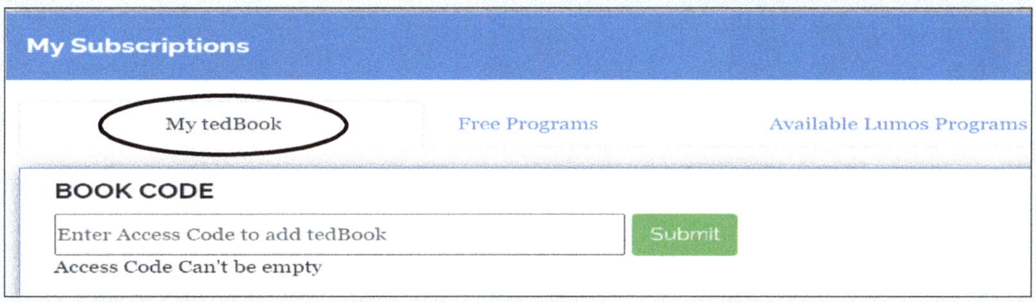

Step 3 ⟶ **Add the new book**

To add the new book for a registered student, choose the '**Student**' button and click on submit.

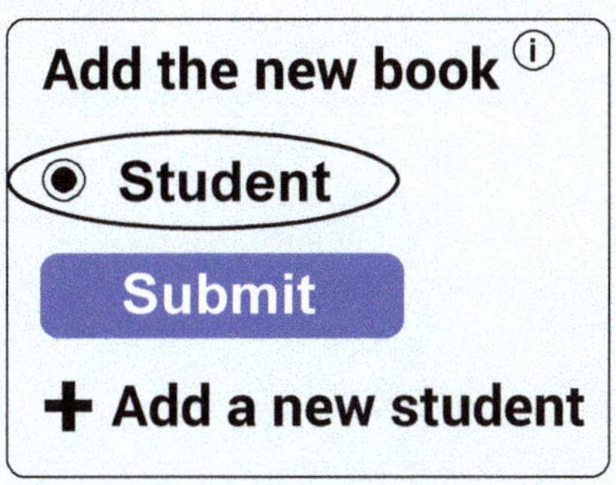

To add the new book for a new student, choose the '**Add New Student**' button and complete the student registration.

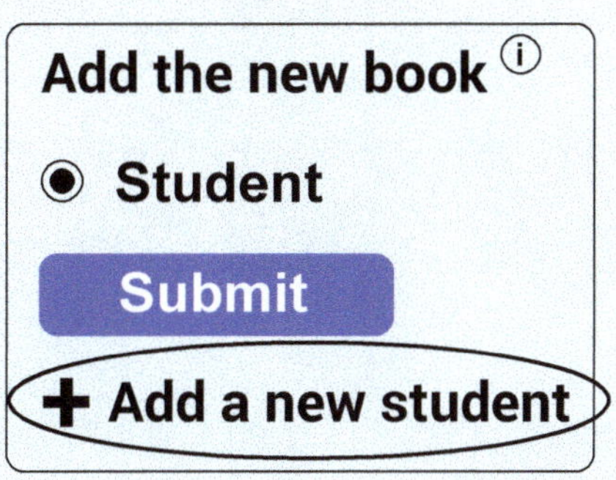

www.ingramcontent.com/pod-product-compliance
Lightning Source LLC
LaVergne TN
LVHW070855210926
848482LV00016B/939